Healthy Body
Making Good Decisions

by Kate Boehm Jerome

Table of Contents

Develop Language 2

CHAPTER 1   Take Charge 4
                         Your Turn: Communicate 9

CHAPTER 2   How Drugs Affect
                         Body Systems 10
                         Your Turn: Summarize 15

CHAPTER 3   Staying Safe 16
                         Your Turn: Interpret Data 19

Career Explorations 20
Use Language to Compare 21
Science Around You 22
Key Words 23
Index 24

Millmark
EDUCATION

DEVELOP LANGUAGE

How can you keep your body healthy? First, you have to understand what a healthy body needs. Then you have to make good decisions, or choices.

Good decisions help your body get what it needs to stay healthy. Good decisions can also protect your body from **disease** and injury, or harm.

Discuss the pictures on these pages. Ask and answer questions like these:

Why is it important to wash your hands before you eat?

What good decisions are the people around the table making?

Why is exercise important?

Describe some decisions that you made that helped your body stay healthy.

> **disease** – a condition that causes illness

washing hands

skateboarding

2 Healthy Body: Making Good Decisions

CHAPTER 1

Take Charge

Fruits and vegetables contain many different nutrients, such as water, vitamins, and minerals.

What are you going to have for lunch? Choosing what foods to eat is one of the most important decisions you make each day.

Food contains **nutrients** that give your body energy and other things it needs. The most important nutrient you need each day is water. You can get water by drinking it and by eating foods that contain a lot of water, such as fruits.

Vitamins and **minerals** are other nutrients that help your body grow and work properly. Fruits and vegetables contain many different vitamins and minerals.

nutrients – substances the body needs to get energy or make the body work

vitamins – one of the groups of nutrients the body needs to be healthy

minerals – substances formed in nature; some are nutrients the body needs

Certain nutrients supply energy to your body. Carbohydrates provide the fastest source of energy. Before a sports activity, many athletes eat meals that have lots of carbohydrates. This gives them the energy they need to perform.

Proteins are needed to build and repair your body. For example, the muscles in your body are made of proteins. Extra proteins can be used for energy.

Fats also provide energy. They carry certain vitamins throughout your body. We need to eat fats every day, but we do not need to eat as much of this nutrient as others.

Nutrients that Provide Energy

carbohydrates  Foods such as pasta, tortillas, vegetables, and fruits contain carbohydrates.

proteins  Foods such as meat, cheese, and peanut butter contain proteins.

fats  Foods such as butter, whole milk, and nuts contain fats.

KEY IDEA Water, vitamins, minerals, carbohydrates, proteins, and fats are important nutrients your body needs each day.

Choose Carefully!

Different foods contain different nutrients. This means you have to eat a wide variety of foods to get all the nutrients you need. But there are so many different choices! How do you know which foods contain the nutrients you need?

The **food pyramid** is a useful guide. It shows different food groups. If you eat foods from each of these groups every day, there is a good chance you will get the important nutrients you need.

> **food pyramid** – a guide to help people make healthy food choices every day
>
> **calories** – units that measure the amount of energy in foods

The Food Pyramid

GRAINS | VEGETABLES | FRUITS | OILS | MILK | MEAT & BEANS

By The Way...

Calories measure the amount of energy in foods. Some foods, such as cookies and candies, contain few nutrients but have many calories. If you take in more calories than your body needs, the extra energy is stored as fat in your body.

Keep Moving!

The food pyramid also reminds you that your body needs exercise every day. Exercise is any physical activity that improves health. You do not have to be on a sports team to get exercise. You can get exercise from just walking or biking. The important thing is to get moving for at least an hour each day.

Different movements help your body in different ways. **Aerobic exercise** makes your body use more oxygen. This strengthens your heart and your lungs. Strength-building exercise helps build muscle.

aerobic exercise – a physical activity that makes the body use more oxygen

Explore Language

GREEK WORD ROOTS
aerobic
aero- = of the air

SHARE IDEAS Tell a friend what kind of exercise you get each day.

Chapter 1: Take Charge

Get Some Sleep

Sometimes it is hard to sleep when there is so much to do! However, making time for sleep is another important decision that can lead to good health.

Your body rests and repairs itself during sleep. A chemical that controls growth is also released during sleep. Scientists also think sleep helps our brain organize and store information.

The amount of sleep you need depends upon your age and activity level. You need less sleep now than you did when you were a baby. But you still probably need at least 9 hours of sleep to feel your best.

KEY IDEA Deciding to eat healthy foods, exercise, and get enough sleep are good choices that lead to good health.

Most babies need 14 or 15 hours of sleep every 24 hours.

Older children need 10 or 11 hours of sleep.

YOUR TURN

COMMUNICATE

Make a chart like the one below. Under each group, list examples of foods you eat. Use the food pyramid on page 6 to help you. Describe your list to a friend. How are your lists different and the same?

Grains	Vegetables	Fruits	Milk	Meat & Beans

MAKE CONNECTIONS

Jumping rope is both an aerobic and strength-building exercise. It makes your body use more oxygen, and it builds muscles in your arms and legs. What is another activity that includes both kinds of exercise? Explain your answer.

USE THE LANGUAGE OF SCIENCE

What are the main kinds of nutrients your body needs every day?

The main nutrients are water, vitamins, minerals, carbohydrates, proteins, and fats.

Chapter 1: Take Charge

CHAPTER 2

How Drugs Affect Body Systems

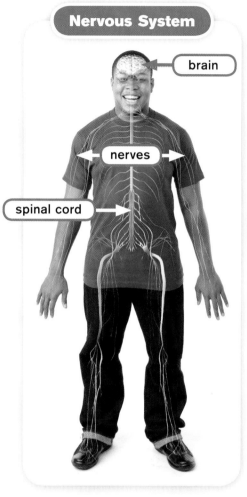

Your body is made up of many **body systems**. A body system is made up of groups of **organs** that do certain things. Each organ also works with other organs to allow the body system to do its main job.

For example, your brain is an organ in your **nervous system**. But your brain needs information from other organs, such as your eyes and ears. All the organs of the nervous system are connected by nerves. Together, all these organs allow the nervous system to do its main job of controlling your body.

body systems – groups of organs that work together to do a job

organs – body parts that do things to help a body system work

nervous system – the body system that controls your body

The organs of your body are made up of groups of special **tissues** that work together. For example, brain tissue makes up your brain.

Tissue is made up of groups of **cells**, which are the smallest units of life. Cells are so tiny that they cannot be seen without a microscope.

Activities in your cells keep your body working. **Drugs** can affect these activities. In this way, drugs can affect body systems.

> **tissues** – groups of cells that work together to do a job
> **cells** – the smallest units of life
> **drugs** – chemicals that affect the way the body works

What Makes Up a Body System?

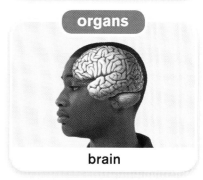
organs — brain

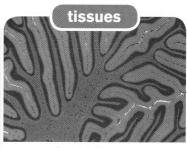
tissues — brain tissue

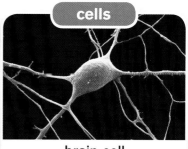
cells — brain cell

KEY IDEA Body systems are made of groups of organs. Organs are made of groups of tissues. Tissues are made of groups of cells.

Some Other Major Body Systems	
System	**Main Job**
circulatory system	moves blood
skeletal system	supports the body
digestive system	breaks down food
muscular system	moves the body
immune system	protects against disease

Chapter 2: How Drugs Affect Body Systems

Helpful Drugs

As you know, sometimes people get sick. A **communicable disease**, such as a cold, is an illness that is spread by germs, or **pathogens**. A **noncommunicable disease**, such as an allergy, cannot be spread.

Doctors carefully use helpful drugs, called **medicines**, to prevent and treat diseases. For example, a **vaccine** is a medicine that can protect against a communicable disease, such as measles. Helpful drugs can improve how body systems work. Many helpful medicines save millions of lives each year.

▲ **Talking with a health professional about helpful medicines is a good decision.**

communicable disease – an illness that can be spread

pathogens – germs that can cause diseases

noncommunicable disease – an illness that cannot be spread

medicines – helpful drugs

vaccine – a medicine that can help protect your body against communicable diseases

KEY IDEA Medicines are helpful drugs that can be used to treat some diseases.

Harmful Drugs

Harmful drugs are not part of a healthy lifestyle. For example, **alcohol** is a drug that slows down the actions of the nervous system. This means signals cannot move as quickly to and from the brain. When people overuse alcohol, their bodies do not move properly and their reaction time is slowed down.

If a person drinks too much alcohol, alcohol poisoning can occur. This can cause the nervous system to completely shut down. A person may not be able to breathe. Without help, death will follow.

alcohol – a drug that slows down the actions of the nervous system and can be harmful

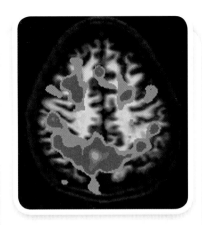

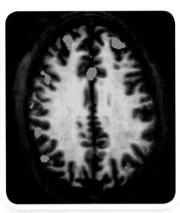

The pink areas in these images show brain activity. The top image shows normal brain activity. The bottom image shows reduced brain activity because of alcohol.

Each year in the United States, about 16,000 people are killed in crashes that involve alcohol.

Chapter 2: How Drugs Affect Body Systems

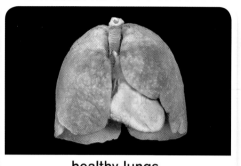
healthy lungs

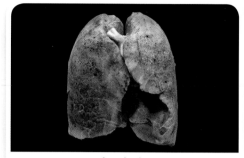
smoker's lungs

▲ The lungs are organs in the respiratory system that allow you to breathe. Smoking can damage the lungs.

Harmful Drugs Can Be Addictive

Once a person starts smoking or using other tobacco products, it is very hard to quit! That is because tobacco contains a drug called **nicotine**.

Nicotine, like alcohol and other harmful drugs, can be **addictive**. An addictive drug can make a person's mind and body depend upon the drug. This means people want to keep using the drug even if the drug is harming their body systems.

If you have a problem with an addictive drug, or need help because someone you know has a problem, you should talk to a trusted adult right away. Parents, school counselors, and churches are good places to go for information.

nicotine – an addictive drug found in tobacco products

addictive – causing the body to develop a strong need for something

KEY IDEAS Harmful drugs can be addictive. Choosing to avoid alcohol and other harmful drugs is a healthy decision.

YOUR TURN

SUMMARIZE

Draw a chart like the one show below.
Summarize how drugs can be both helpful and harmful.

	What They Do	Example
Helpful Drugs		
Harmful Drugs		

MAKE CONNECTIONS

What places are available in your community to help people with drug problems?

Monitor Comprehension

What did you find hard to understand in this chapter?
What strategies did you use to understand it better?

Chapter 2: How Drugs Affect Body Systems

CHAPTER 3

Staying Safe

It is fun to play on a sports team. It is also good for your body systems to exercise and stay active. However, any activity you do has some **risk**, or chance of injury.

Does this mean you should avoid being active? No, it just means that you need to make smart decisions to protect yourself. For example, if you play a sport that requires safety equipment, you should wear it whenever you play.

Basic Sports Safety

1. Wear proper safety equipment, such as helmets and pads.
2. Follow the rules.
3. Look out for others.
4. Do not play if you are injured.

risk – the chance of injury or danger

SHARE IDEAS Look at the photo. **Describe** the safety equipment you see. How does this equipment help protect the player?

Healthy Body: Making Good Decisions

Each year, tens of thousands of kids in the United States are treated for injuries they receive in bike accidents. Most of these accidents involve head injuries.

You may not like to wear a bike helmet. But make a smart decision and wear one anyway! A helmet can help protect your brain from injury if you have an accident.

A bike path is safer than a city street. However, even on a bike path, you should obey traffic signals. Stop at the stop signs, and use hand signals when you make a turn.

▲ Always ride on the right in the same direction as traffic. Never ride against traffic.

▲ Hand signals help you stay safe on a bike.

KEY IDEA Wearing proper safety equipment and following rules can reduce the risk of serious injury.

Chapter 3: Staying Safe

Make a Plan

In an **emergency**, you have to think and act fast to stay safe. So it is a good idea to make a plan for what you might do before an emergency happens.

▼ Having escape plans can keep you safe in a fire at home or at school.

You have probably been in a fire drill at school. This kind of drill allows you to practice what you should do and where you should go if a fire breaks out. Some schools also have tornado and earthquake drills. You should think about how you would keep yourself safe if any of these emergencies happen in other places, too.

In an emergency, it is important to stay calm. This will help you make good decisions that can keep you healthy and safe.

emergency – a sudden need for fast action

KEY IDEA Making a plan before an emergency happens can help you keep safe.

YOUR TURN

INTERPRET DATA

The chart shows the number of head injuries that occurred in different activities in one year. Interpret the data to answer the questions.

1. Did more people have head injuries due to playing baseball or football?

2. What activity caused the highest number of head injuries?

3. Why is it important to wear a helmet while doing these activites?

Activity	Estimated Number of Head Injuries
bicycles	151,024
skateboards	18,743
snowboarding	8,540
football	51,953
ice hockey	5,944
baseball	63,234

Source: Consumer Product Safety Commission

MAKE CONNECTIONS

What rules does your school have that help keep you safe and healthy? Make a list and share it.

EXPAND VOCABULARY

You have read about safety equipment. The noun **equipment** is related to the verb **equip**. Write sentences that show the meanings of these words. Then write sentences for these noun/verb pairs:

pavement/pave   **argument/argue**   **government/govern**

Chapter 3: Staying Safe

CAREER EXPLORATIONS

Guidance Counselors: What Do They Do?

High school guidance counselors are good at giving advice. They help students make good career choices. Guidance counselors can also help students deal with problems.

Guidance Counselor Log

8 a.m.	Meet with Simone C. about college plans.
9 a.m.	Meet with D. L. about behavior problems in class. Discuss ways to help and schedule follow-up meetings.
10 a.m.	Work on class schedules for next year's freshmen.
11 a.m.	Meet with the parents of Lucas S. to discuss his progress in school.
1 p.m.	Give career survey test to Grade 10 students.
2 p.m.	Write letter of recommendation for Lupe V.

- Read the guidance counselor's log.

- Tell what you would like and would not like about this career.

USE LANGUAGE TO COMPARE

Use Sentence Patterns

When you compare, you tell how things are similar and different. Some sentence patterns can help you make comparisons.

EXAMPLES

Both walking **and** swimming **are** good forms of exercise.

Grapes **are healthier than** candy.

Rice **has more nutrients than** potato chips.

With a friend, look at the photos in this book. Make comparisons about what you see. Use sentence patterns like the ones above.

Write a Comparison

You have learned about nutrients and the food pyramid. Describe two meals and compare them. Tell which meal is healthier.

- Decide whether you will describe two breakfasts, lunches, or dinners.
- Describe each meal.
- Then compare the meals. Use sentence patterns to explain how they are similar and different.

Words You Can Use

both	healthier than
the same as	more than
different from	less than

Use Language to Compare

SCIENCE AROUND YOU

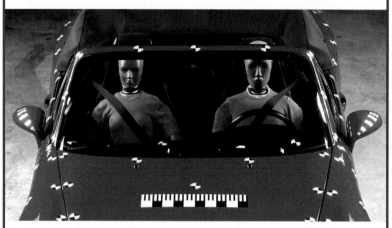

Don't Be a Dummy! Use Your Seat Belt!

Scientists use crash test dummies to test safety equipment in cars. These dummies are built like human bodies. Special sensors tell scientists how the dummy bodies are affected in a crash. This helps scientists decide how to design, or make, better safety equipment.

Crash dummies test seat belts and air bags, which save thousands of lives each year.

Read the poster and answer the questions.

- Why do scientists build crash test dummies like human bodies?

- What do scientists do with information they collect from crash test dummies?

Key Words

addictive causing the body to develop a strong need for something
Harmful drugs can be **addictive**.

aerobic exercise (aerobic exercises) a physical activity that makes the body use more oxygen
Running is an **aerobic exercise**.

body system (body systems) a group of organs that work together to do a job
The nervous system is a **body system** that controls your body.

cell (cells) the smallest unit of life
A **cell** is the building block of all body systems.

communicable disease (communicable diseases) an illness that can be spread
Measles are a **communicable disease** that can pass from one person to another.

drug (drugs) a chemical that affects the way the body works
Medicines are **drugs** that can help our body systems.

food pyramid a guide to help people make healthy food choices every day
The **food pyramid** shows healthy groups of foods to eat and reminds us to exercise.

noncommunicable disease (noncommunicable diseases) an illness that cannot be spread
An allergy is a **noncommunicable disease**.

nutrient (nutrients) a substance the body needs to get energy or make the body work
Water is a very important **nutrient** that our body needs to survive.

organ (organs) a body part that does things to help a body system work; a group of tissues that work together to do a job
The brain is an important **organ** in the nervous system.

pathogen (pathogens) a germ that can cause disease
Washing your hands can destroy some **pathogens** that cause disease.

tissue (tissues) a group of cells that work together to do a job
Groups of brain cells form brain **tissue**.

Index

addictive 14
aerobic exercise 7, 9
alcohol 13–14
alcohol poisoning 13
body system 10–12, 14, 16
calorie 6
carbohydrate 5, 9
cell 11
communicable disease 12
decision 2–3, 4, 8, 12, 14, 16–18

disease 2–3, 11, 12
drug 11–14, 15
emergency 18
energy 4–5
exercise 2–3, 7–8, 9, 16, 21
fat 5, 9
food pyramid 6–7, 9, 21
medicine 12
microscope 11
mineral 4–5, 9

nervous system 10, 13
nicotine 14
nutrient 4–5, 6, 9, 21
organ 10–11, 14
pathogen 12
protein 5, 9
risk 16
sleep 8
tissue 11
vaccine 12
vitamin 4–5, 9

MILLMARK EDUCATION CORPORATION
Ericka Markman, President and CEO; Karen Peratt, VP, Editorial Director; Lisa Bingen, VP, Marketing; Rachel L. Moir, Director, Operations and Production; Shelby Alinsky, Assistant Editor; Mary Ann Mortellaro, Science Editor; Kris Hanneman, Photo Research

PROGRAM AUTHORS
Mary Hawley; Program Author, Instructional Design
Kate Boehm Jerome; Program Author, Science

BOOK DESIGN Steve Curtis Design

CONTENT REVIEWER
Kefyn M. Catley, PhD, Western Carolina University, Cullowhee, NC

PROGRAM ADVISORS
Scott K. Baker, PhD, Pacific Institutes for Research, Eugene, OR
Carla C. Johnson, EdD, University of Toledo, Toledo, OH
Donna Ogle, EdD, National-Louis University, Chicago, IL
Betty Ansin Smallwood, PhD, Center for Applied Linguistics, Washington, DC
Gail Thompson, PhD, Claremont Graduate University, Claremont, CA
Emma Violand-Sánchez, EdD, Arlington Public Schools, Arlington, VA (retired)

TECHNOLOGY
Arleen Nakama, Project Manager
Audio CDs: Heartworks International, Inc.
CD-ROMs: Cannery Agency

PHOTO CREDITS cover ©CORBIS/agefotostock; IFC and 15b ©David Safanda/iStockphoto.com; 1 ©Aurora Creative/Getty Images; 2a ©AJPhoto/Photo Researchers, Inc.; 2b, 8a, 23 ©David Young-Wolff/PhotoEdit; 2-3 ©Jim Lane/Alamy; 3a ©Ryan McVay/Photodisc/Getty Images; 3b ©Polka Dot Images/Photolibrary; 4 ©Ilene MacDonald/Alamy; 5a ©Feng Yu/Shutterstock; 5b ©LockStockBob/Shutterstock; 5c ©Harris Shiffman/Shutterstock; 6 illustration by Dino Idrizbegovic; 7 ©Steve Skjold/Alamy; 8b ©Goodshoot/Corbis; 9a, 12, 17b, 17c, 17d, 18a, 18b ©Michael Newman/PhotoEdit; 9b and 9c photos by Ken Karp for Millmark Education; 10 and 11a photos by Ken Karp for Millmark Education, illustrations by Craig Bowman; 11a Photo by Ken Karp for Millmark Education, Illustration by Craig Bowman; 11b ©Innerspace Imaging/Photo Researchers, Inc.; 11c ©David McCarthy/Photo Researchers, Inc.; 13a and 13b ©Susan F. Tapert, Ph. D., Associate Professor of Psychiatry, University of California San Diego; 13c ©Enigma/Alamy; 14a and 14b ©Ralph Hutchings/Visuals Unlimited; 15a ©Stockbyte Platinum/Getty Images; 16 ©Wesley Hitt/The Image Bank/Getty Images; 17a ©Chris Howes/Wild Places Photography/Alamy; 20 ©Mary Kate Denny/PhotoEdit; 22 MAXIMILIAN STOCK LTD/Oxford Scientific/Photolibrary; 24 ©Matthew T Tourtellott/Shutterstock

Copyright ©2008 Millmark Education Corporation

All rights reserved. Reproduction of the whole or any part of the contents without written permission from the publisher is prohibited. Millmark Education and ConceptLinks are registered trademarks of Millmark Education Corporation.

Published by Millmark Education Corporation
7272 Wisconsin Avenue, Suite 300
Bethesda, MD 20814

ISBN-13: 978-1-4334-0248-7

Printed in the USA

10 9 8 7 6 5 4 3 2 1

24 *Healthy Body: Making Good Decisions*